WILLIAM PARSONS

GREAT ASTRONOMERS

ROBERT STAWELL BALL

GIFT *Certificate*

TO: ______________________

____

FROM: ______________________

____

DATE: ______________________

WILLIAM PARSONS

GREAT ASTRONOMERS

BY

ROBERT STAWELL BALL

D.Sc, LL.D., F.R.S.

(July 1, 1840 – November 25, 1913)

LOWNDEAN PROFESSOR OF ASTRONOMY AND GEOMETRY
IN THE UNIVERSITY OF CAMBRIDGE

DIAMOND BOOKS

www.diamondbooks.ca

TORONTO, CANADA – 2017

DIAMOND BOOKS - CANADA

DIAMOND BOOKS

Toronto, ON, CANADA

http://www.diamondbooks.ca

BIBLIOGRAPHIC INFORMATION

'WILLIAM PARSONS'
Is an extract or a chapter from the book
"GREAT ASTRONOMERS" by Robert Stawell Ball.

FIRST PUBLISHED IN: 1896.

PUBLISHED IN CANADA

Published in Canada by DIAMOND BOOKS - CANADA, an imprint of DIAMOND PUBLISHERS - http://www.diamondpublishers.com

FIRST EDITION: APRIL, 2017.

PAPERBACK EDITION : ISBN: 978-1-988357-58-4

PRINTED IN CANADA

PREFACE
TO THE ORIGINAL EDITION

It has been my object in these pages to present the life of each astronomer in such detail as to enable the reader to realise in some degree the man's character and surroundings; and I have endeavoured to indicate as clearly as circumstances would permit the main features of the discoveries by which he has become known.

There are many types of astronomers—from the stargazer who merely watches the heavens, to the abstract mathematician who merely works at his desk; it has, consequently, been necessary in the case of some lives to adopt a very different treatment from that which seemed suitable for others.

ROBERT STAWELL BALL.

The Observatory, Cambridge.
October, 1895.

NOTE: The full-length of 'Preface' can be found in its original edition of the book "GREAT ASTRONOMERS".

- ROBERT STAWELL BALL

INTRODUCTION

Of all the natural sciences there is not one which offers such sublime objects to the attention of the inquirer as does the science of astronomy. From the earliest ages the study of the stars has exercised the same fascination as it possesses at the present day. Among the most primitive peoples, the movements of the sun, the moon, and the stars commanded attention from their supposed influence on human affairs. The practical utilities of astronomy were also obvious in primeval times. Maxims of extreme antiquity show how the avocations of the husbandman are to be guided by the movements of the heavenly bodies. The positions of the stars indicated the time to plough, and the time to sow. To the mariner who was seeking a way across the trackless ocean, the heavenly bodies .offered the only reliable marks by which his path could be guided. There was, accordingly, a stimulus both from intellectual curiosity and from practical necessity to follow the movements of the stars. Thus began a search for the causes of the ever-varying phenomena which the heavens display.

- ROBERT STAWELL BALL

Many of the earliest discoveries are indeed prehistoric. The great diurnal movement of the heavens, and the annual revolution of the sun, seem to have been known in times far more ancient than those to which any human monuments can be referred. The acuteness of the early observers enabled them to single out the more important of the wanderers which we now call planets. They saw that the star-like objects, Jupiter, Saturn, and Mars, with the more conspicuous Venus, constituted a class of bodies wholly distinct from the fixed stars among which their movements lay, and to which they bear such a superficial resemblance. But the penetration of the early astronomers went even further, for they recognized that Mercury also belongs to the same group, though this particular object is seen so rarely. It would seem that eclipses and other phenomena were observed at Babylon from a very remote period, while the most ancient records of celestial observations that we possess are to be found in the Chinese annals. The study of astronomy, in the sense in which we understand the word, may be said to have commenced under the reign of the Ptolemies at Alexandria. The most famous name in the science of this period is that of Hipparchus, who lived and worked at Rhodes about the year 160 B.C. It was his splendid investigations that first wrought the observed facts into a coherent branch of knowledge. He recognized the primary obligation which lies on the student of the heavens to compile as complete an inventory as possible of the objects which are there to be found. Hipparchus accordingly commenced by undertaking, on a small scale, a task exactly similar to that on which modern astronomers, with all available appliances of meridian circles, and photographic telescopes, are constantly engaged at the present day. He compiled a catalogue of the principal fixed stars, which is of special value to astronomers, as being the earliest work of its kind which has been handed down. He also studied the movements of the sun and the moon, and framed theories to account for the incessant changes which he saw in progress. He found a much more difficult problem in his attempt to interpret satisfactorily the complicated movements of the planets. With the view of constructing a theory which should give some coherent account of the subject, he made many observations of the places of these wandering stars. How great were the advances which Hipparchus accomplished may be appreciated if we reflect that, as a preliminary task to his more purely astronomical labours, he had to invent that branch of mathematical science by which alone the problems he proposed could be solved. It was for this purpose that he devised the indispensable method of calculation which we now know so well as trigonometry. Without the aid rendered by this beautiful art it would have been impossible for any n ally important advance in astronomical calculation to have been effected.

But the discovery which shows, beyond all others, that Hipparchus possessed one of the master-minds of all time was the detection of that remarkable celestial movement known as the precession of the equinoxes. The inquiry which conducted to this discovery involved a most profound investigation, especially when it is remembered that in the days of Hipparchus the means of observation of the heavenly bodies were only of the rudest description, and the available observations of earlier dates were extremely scanty. We can but look with astonishment on the genius of the man who, in spite of such difficulties, was able to detect such a phenomenon as the precession, and to exhibit its actual magnitude. I shall endeavour to explain the nature of this singular celestial movement, for it may be said to offer the first instancc in the history of science in which we find that combination of accurate observation with skilful interpretation, of which, in the subsequent development of astronomy, we have so many splendid examples.

The word equinox implies the condition that the night is equal to the day. To a resident on the equator the night is no doubt equal to the day at all times in the year, but to one who lives on any other part of the earth, in either hemisphere, the night and the day are not generally equal. There is, however, one occasion in spring, and another in autumn, on which the day and the night are each twelve hours at all places on the earth. When the night and day are equal in spring, the point which the sun occupies on. the heavens is termed the vernal equinox. There is similarly another point in which the sun is situated at the time of the autumnal equinox. In any investigation of the celestial movements the positions of these two equinoxes on the heavens are of primary importance, and Hipparchus, with the instinct of genius, perceived their significance, and commenced to study them. It will be understood that we can always define the position of a point on the sky with reference to the surrounding stars. No doubt we do not see the stars near the sun when the sun is shining, but they are there nevertheless. The ingenuity of Hipparchus enabled him to determine the positions of each of the two equinoxes relatively to the stars which lie in its immediate vicinity. After examination of the celestial places of these points at different periods, he was led to the conclusion that each equinox was moving relatively to the stars, though that movement was so slow that twenty five thousand years would necessarily elapse before a complete circuit of the heavens was accomplished. Hipparchus traced out this phenomenon, and established it on an impregnable basis, so that all astronomers have ever since recognised the precession of the equinoxes) as one of the fundamental facts of astronomy. Not until nearly two thousand years after Hipparchus had made this splendid discovery was the explanation of its cause given by Newton.

From the days of Hipparchus down to the present hour the science of astronomy has steadily grown. One great observer after another has appeared from time to time, to reveal some new phenomenon with regard to the celestial bodies or their movements, while from time to time one commanding intellect after another has arisen to explain the true import of the facts of observations. The history of astronomy thus becomes inseparable from the history of the great men to whose labours its development is due. In the ensuing chapters we have endeavoured to sketch the lives and the work of the great philosophers, by whose labours the science of astronomy has been created.

We shall commence with Ptolemy, who, after the foundations of the science had been laid by Hipparchus, gave to astronomy the form in which it was taught throughout the Middle Ages. We shall next see the mighty revolution in our conceptions of the universe which are associated with the name of Copernicus. We then pass to those periods illumined by the genius of Galileo and Newton, and afterwards we shall trace the careers of other more recent discoverers, by whose industry and genius the boundaries of human knowledge have been so greatly extended. Our history will be brought down late enough to include some of the illustrious astronomers who laboured in the generation which has just passed away.

- ROBERT STAWELL BALL

TABLE OF CONTENTS

CHAPTERS

**An original Chapter from the book: "GREAT ASTRONOMERS", by Robert Stawell Ball.*

CHAPTER

14

GREAT ASTRONOMERS

The Earl of Rosse

(William Parsons, 3rd Earl of Rosse)

(June 17, 1800 – October 31, 1867)

- ROBERT STAWELL BALL

William Parsons, 3rd Earl of Rosse
(June 17, 1800 – October 31, 1867)

CHAPTER 14

GREAT ASTRONOMERS: WILLIAM PARSONS (3RD EARL OF ROSSE)

THE subject of our present sketch occupies quite a distinct position in scientific history. Unlike many others who have risen by their scientific discoveries from obscurity to fame, the great Earl of Rosse was himself born in the purple.1 His father, who, under the title of Sir Lawrence Parsons, had occupied a distinguished position in the Irish Parliament, succeeded on the death of his father to the Earldom which had been recently created. The subject of our present memoir was, therefore, the third of the

Earls of Rosse.2 He was born in York on June 17, 1800. Prior to his father's death in 1841, he was known as Lord Oxmantown.

The University education of the illustrious astronomer was begun in Dublin and completed at Oxford. We do not hear in his case of any very remarkable University career. Lord Rosse was, however, a diligent student, and obtained a first-class in mathematics. He always took a great deal of interest in social questions, and was a profound student of political economy. He had a seat in the House of Commons, as member for King's County, from 1821 to 1834, his ancestral estate being situated in this part of Ireland.

Lord Rosse was endowed by nature with a special taste for mechanical pursuits. Not only had he the qualifications of a scientific engineer, but he had the manual dexterity which qualified him personally to carry out many practical arts. Lord Rosse was, in fact, a skillful mechanic, an experienced founder, and an ingenious optician. His acquaintances were largely among those who were interested in mechanical pursuits, and it was his delight to visit the works or engineering establishments where refined processes in the arts were being carried on. It has often been stated—and as I have been told by members of his family, truly stated—that on one occasion, after he had been shown over some large works in the north of England, the proprietor bluntly said that he was greatly in want of a foreman, and would indeed be pleased if his visitor, who had evinced such extraordinary capacity for mechanical operations, would accept the post. Lord Rosse produced his card, and gently explained that he was not exactly the right man, but he appreciated the compliment, and this led to a pleasant dinner, and was the basis of a long friendship.

- ROBERT STAWELL BALL

I remember on one occasion hearing Lord Rosse explain how it was that he came to devote his attention to astronomy. It appears that when he found himself in the possession of leisure and of means, he deliberately cast around to think how that means and that leisure could be most usefully employed. Nor was it surprising that he should search for a direction which would offer special scope for his mechanical tastes. He came to the conclusion that the building of great telescopes was an art which had received no substantial advance since the great days of William Herschel. He saw that to construct mighty instruments for studying the heavens required at once the command of time and the command of wealth, while he also felt that this was a subject the inherent difficulties of which would tax to the uttermost whatever mechanical skill he might possess. Thus it was he decided that the construction of great telescopes should become the business of his life.

Birr Castle

In the centre of Ireland, seventy miles from Dublin, on the border between King's County and Tipperary, is a little town whereof we must be cautious before writing the name. The inhabitants of that town frequently insist that its name is **Birr**,* while the official designation is Parsonstown, and to this day for every six people who apply one name to the town, there will be half a dozen who use the other. But whichever it may be, Birr or Parsonstown—and I shall generally call it by the latter name—it is a favourable specimen of an Irish county town. The widest street is called the Oxmantown Mall It is bordered by the dwelling-houses of the chief residents, and adorned with rows of stately trees. At one end of this distinctly good feature in the town is the Parish Church, while at the opposite end are the gates leading into Birr Castle, the ancestral home of the house of Parsons.

* Considering the fame acquired by Parsonstown from Lord Rosse's mirrors, it may be interesting to note the following extract from "The Natural History of Ireland," by Dr. Gerard Boate, Thomas Molinenx, M.D., F.R.S., and others, which shows that 150 years ago Parsonstown was famous for its glass:— "We shall conclude this chapter with the glass, there having been several glasshouses set up by the English in Ireland, none in Dublin or other cities, but all of them in the country; amongst which the principal was that of Birre, a market town, otherwise called Parsonstown, after one Sir Lawrence Parsons, who, having purchased that lordship, built a goodly house upon it; his son" William Parsons having succeeded him in the possession of it; which town is situate in Queen's County, about fifty miles (Irish) to the south-west of Dublin, upon the borders of the two provinces of Leinster and Munster; from this place Dublin was furnished with all sorts of window and drinking glasses, and such other as commonly are in use. One part of the materials, viz., the sand, they had out of England; the other, to wit the ashes, they made in the place of ash-tree, and used no other. The chiefest difficulty was to get the clay for the pots to melt the materials in; this they had out of the north."—Chap. XXI., Sect. VIII., "Of the Glass made in Ireland."

The Mall. Parsonstown.

Passing through the gates the visitor enters a spacious demesne, possessing much beauty of wood and water, one of the most pleasing features being the junction of the two rivers, which unite at a spot ornamented by beautiful timber. At various points illustrations of the engineering skill of the great Earl will be observed. The beauty of the park has been greatly enhanced by the construction of an ample lake, designed with the consummate art by which art is concealed. Even in mid-summer it is enlivened by troops of wild ducks preening themselves in that confidence which they enjoy in those happy localities where the sound of a gun is seldom heard. The water is led into the lake by a tube which passes under one of the two rivers just mentioned, while the overflow from the lake turns a water-wheel, which works a pair of elevators ingeniously constructed for draining the low-lying parts of the estate.

- ROBERT STAWELL BALL

Birr Castle itself is a noble mansion with reminiscences from the time of Cromwell. It is surrounded by a moat and a drawbridge of modern construction, and from its windows beautiful views can be had over the varied features of the park. But while the visitors to Parsons-town will look with great interest on this residence of an Irish landlord, whose delight it was to dwell in his own country, and among his own people, yet the feature which they have specially come to observe is not to be found in the castle itself. On an extensive lawn, sweeping down from the moat towards the lake, stand two noble masonry walls. They are turreted and clad with ivy, and considerably loftier than any ordinary house. As the visitor approaches, he will see between those walls what may at first sight appear to him to be the funnel of a steamer lying down horizontally. On closer approach he will find that it is an immense wooden tube, sixty feet long, and upwards of six feet in diameter. It is in fact large enough to admit of a tall man entering into it and walking erect right through from one end to the other. This is indeed the most gigantic instrument which has ever been constructed for the purpose of exploring the heavens. Closely adjoining the walls between which the great tube swings, is a little building called “The Observatory. “In this the smaller instruments are contained, and there are kept the books which are necessary for reference. The observatory also offers shelter to the observers, and provides the bright fire and the cup of warm tea, which are so acceptable in the occasional intervals of a night's observation passed on the top of the walls with no canopy but the winter sky.

Almost the first point which would strike the visitor to Lord Rosse's telescope is that the instrument at which he is looking is not only enormously greater than anything of the kind that he has ever seen before, but also that it is something of a totally different nature. In an ordinary telescope he is accustomed to find a tube with lenses of glass at either end, while the large telescopes that we see in our observatories are also in general constructed on the same principle. At one end there is the object-glass, and at the other end the eye-piece, and of course it is obvious that with an instrument of this construction it is to the lower end of the tube that the eye of the observer must be placed when the telescope is pointed to the skies. But in Lord Rosse's telescope you would look in vain for these glasses, and it is not at the lower end of the instrument that you are to take your station when you are going to make your observations. The astronomer at Parsonstown has rather to avail himself of the ingenious system of staircases and galleries, by which he is enabled to obtain access to the mouth of the great tube. The colossal telescope which swings between the great walls, like Herschel's great telescope already mentioned, is a reflector, the original invention of which is due of course to Newton. The optical work which is accomplished by the lenses in the ordinary telescope is effected in the type of instrument constructed by Lord Rosse by a reflecting mirror which is placed at the lower end of the vast tube. The mirror in this instrument is made of a metal consisting of two parts of copper to one of tin. As we have already seen, this mixture forms an alloy of a very peculiar nature. The copper and the tin both surrender their distinctive qualities, and unite to form a material of a very different physical character. The copper is tough and brown, the tin is no doubt silvery in hue, but soft and almost fibrous in texture. When the two metals are mixed together in the proportions I have stated, the alloy obtained is intensely hard and quite brittle, being in both these respects utterly unlike either of the two ingredients of which it is composed. It does, however, resemble the tin in its whiteness, but it acquires a lustre far brighter than tin; in fact, this alloy hardly falls short of silver itself in its brilliance when polished.

- ROBERT STAWELL BALL

The first duty that Lord Rosse had to undertake was the construction of this tremendous mirror, six feet across, and about four or five inches thick. The dimensions were far in excess of those which had been contemplated in any previous attempt of the same kind. Herschel had no doubt fashioned one mirror of four feet in diameter, and many others of smaller dimensions, but the processes which he employed had never been fully published, and it was obvious that, with a large increase in dimensions, great additional difficulties had to be encountered. Difficulties began at the very commencement of the process, and were experienced in one form or another at every subsequent stage. In the first place, the mere casting of a great disc of this mixture of tin and copper, weighing something like three or four tons, involved very troublesome problems. No doubt a casting of this size, if the material had been, for example, iron, would have offered no difficulties beyond those with which every practical founder is well acquainted, and which he has to encounter daily in the course of his ordinary work. But speculum metal is a material of a very intractable description.

Lord Rosse's Telescope.
(From a photograph by W. Lawrence, Upper Sackville Street, Dublin.)

There is, of course, no practical difficulty in melting the copper, nor in adding the proper proportion of tin when the copper has been melted. There may be no great difficulty in arranging an organisation by which several crucibles, filled with the molten material, shall be poured simultaneously so as to obtain the requisite mass of metal, but from this point the difficulties begin. For speculum metal when cold is excessively brittle, and were the casting permitted to cool like an ordinary copper or iron casting, the mirror would inevitably fly into pieces. Lord Rosse, therefore, found it necessary to anneal the casting with extreme care by allowing it to cool very slowly. This was accomplished by drawing the disc of metal as soon as it had entered into the solid state, though still glowing red, into an annealing oven. There the temperature was allowed to subside so gradually, that six weeks elapsed before the mirror had reached the temperature of the external air. The necessity for extreme precaution in the operation of annealing will be manifest if we reflect on one of the accidents which happened. On a certain occasion, after the cooling of a great casting had been completed, it was found, on withdrawing the speculum, that it was cracked into two pieces. This mishap was eventually traced to the fact that one of the walls of the oven had only a single brick in its thickness, and that therefore the heat had escaped more easily through that side than through the other sides which were built of double thickness. The speculum had, consequently, not cooled uniformly, and hence the fracture had resulted. Undeterred, however, by this failure, as well as by not a few other difficulties, into a description of which we cannot now enter, Lord Rosse steadily adhered to his self-imposed task, and at last succeeded in casting two perfect discs on which to commence the tedious processes of grinding and polishing. The magnitude of the operations involved may perhaps be appreciated if I mention that the value of the mere copper and tin entering into the composition of each of the mirrors was about £500.

In no part of his undertaking was Lord Rosse's mechanical ingenuity more taxed than in the devising of the mechanism for carrying out the delicate operations of grinding and polishing the mirrors, whose casting we have just mentioned. In the ordinary operations of the telescope-maker, such processes had hitherto been generally effected by hand, but, of course, such methods became impossible when dealing with mirrors which were as large as a good-sized dinner table, and whose weight was measured by tons. The rough grinding was effected by means of a tool of cast iron about the same size as the mirror, which was moved by suitable machinery both backwards and forwards, and round and round, plenty of sand and water being supplied between the mirror and the tool to produce the necessary attrition. As the process proceeded and as the surface became smooth, emery was used instead of sand; and when this stage was complete, the grinding tool was removed and the polishing tool was substituted. The essential part of this was a surface of pitch, which, having been temporarily softened by heat, was then placed on the mirror, and accepted from the mirror the proper form. Rouge was then introduced as the polishing powder, and the operation was continued about nine hours, by which time the great mirror had acquired the appearance of highly polished silver. When completed, the disc of speculum metal was about six feet across and four inches thick. The depression in the centre was about half an inch. Mounted on a little truck, the great speculum was then conveyed to the instrument, to be placed in its receptacle at the bottom of the tube, the length of which was sixty feet, this being the focal distance of the mirror. Another small reflector was inserted in the great tube sideways, so as to direct the gaze of the observer down upon the great reflector. Thus was completed the most colossal instrument for the exploration of the heavens which the art of man had ever constructed.

It was once my privilege to be one of those to whom the illustrious builder of the great telescope entrusted its use. For two seasons in 1865 and 1866 I had the honour of being Lord Rosse's astronomer. During that time I passed many a fine night in the observer's gallery, examining different objects in the heavens with the aid of this remarkable instrument. At the time I was there, the objects principally studied were the nebulae, those faint stains of light which lie on the background of the sky. Lord Rosse's telescope was specially suited for the scrutiny of these objects, inasmuch as their delicacy required all the light-grasping power which could be provided.

Roman Catholic Church at Parsonstown.

One of the greatest discoveries made by Lord Rosse, when his huge instrument was first turned towards the heavens, consisted in the detection of the spiral character of some of the nebulous forms. When the extraordinary structure of these objects was first announced, the discovery was received with some degree of incredulity. Other astronomers looked at the same objects, and when they failed to discern—and they frequently did fail to discern—the spiral structure which Lord Rosse had indicated, they drew the conclusion that this spiral structure did not exist. They thought it must be due possibly to some instrumental defect or to the imagination of the observer. It was, however, hardly possible for any one who was both willing and competent to examine into the evidence, to doubt the reality of Lord Rosse's discoveries. It happens, however, that they have been recently placed beyond all doubt by testimony which it is impossible to gainsay. A witness never influenced by imagination has now come forward, and the infallible photographic plate has justified Lord Rosse.

Among the remarkable discoveries which Dr. Isaac Roberts6 has recently made in the application of his photographic apparatus to the heavens, there is none more striking than that which declares, not only that the nebulae which Lord Rosse described as spirals, actually do possess the character so indicated, but that there are many others of the same description. He has even brought to light the astonishingly interesting fact that there are invisible objects of this class which have never been seen by human eye, but whose spiral character is visible to the peculiar delicacy of the photographic telescope.

In his earlier years, Lord Rosse himself used to be a diligent observer of the heavenly bodies with the great telescope which was completed in the year 1845. But I think that those who knew Lord Rosse well, will agree that it was more the mechanical processes incidental to the making of the telescope which engaged his interest than the actual observations with the telescope when it was completed. Indeed one who was well acquainted with him believed Lord Rosse's special interest in the great telescope ceased when the last nail had been driven into it. But the telescope was never allowed to lie idle, for Lord Rosse always had associated with him some ardent young astronomer, whose delight it was to employ to the uttermost the advantages of his position in exploring the wonders of the sky. Among those who were in this capacity in the early days of the great telescope, I may mention my esteemed friend Dr. Johnston Stoney.

Such was the renown of Lord Rosse himself, brought about by his consummate mechanical genius and his astronomical discoveries, and such the interest which gathered around the marvellous workshops at Birr Castle, wherein his monumental exhibitions of optical skill were constructed, that visitors thronged to see him from all parts of the world. His home at Parsonstown became one of the most remarkable scientific centres in Great Britain; thither assembled from time to time all the leading men of science in the country, as well as many illustrious foreigners. For many years Lord Rosse filled with marked distinction the exalted position of President of the Royal Society, and his advice and experience in practical mechanical matters were always at the disposal of those who sought his assistance.

Personally and socially Lord Rosse endeared himself to all with whom he came in contact. I remember one of the attendants telling me that on one occasion he had the misfortune to let fall and break one of the small mirrors on which Lord Rosse had himself expended many hours of hard personal labour.

The only remark of his lordship was that "accidents will happen."

He was succeeded by his eldest son, the present Earl of Rosse, who has inherited his father's scientific abilities, and done much notable work with the great telescope. His youngest son, the Hon. Charles A. Parsons, is the famous engineer and the inventor of the steam turbine.

COMPLIMENTARY MATERIAL

A BRIEF BIOGRAPHICAL SKETCH

WILLIAM PARSONS

- ROBERT STAWELL BALL

William Parsons, 3rd Earl of Rosse
(June 17, 1800 – October 31, 1867)

COMPLIMENTARY MATERIAL

A BRIEF BIOGRAPHICAL SKETCH

WILLIAM PARSONS

William Parsons, 3rd Earl of Rosse KP PRS (June 17, 1800 – October 31, 1867) was an Anglo-Irish astronomer who had several telescopes built. His 72-inch telescope, built in 1845 and colloquially known as the "Leviathan of Parsonstown", was the world's largest telescope, in terms of aperture size, until the early 20th century. From 1807 until 1841, he was styled as Baron Oxmantown.

Life

He was born in Yorkshire, England, in the city of York, the son of a future Irish peer. He was educated, as Lord Oxmantown, at Trinity College, Dublin, and Oxford University's Magdalen College, graduating with first-class honours in mathematics in 1822. He inherited an earldom and a large estate in King's County (now County Offaly) in Ireland when his father, Lawrence, 2nd Earl of Rosse, died in 1841.

Lord Rosse married Mary Field, daughter of John Wilmer Field, on 14 April 1836. They had a total of thirteen children, but only four sons survived to adulthood:

- Lawrence, 4th Earl of Rosse (known as Baron Oxmantown until 1867; 17 November 1840 – 30 August 1908).

- The Rev. Randal Parsons (26 April 1848 – 15 November 1936).

- The Hon. Richard Clere Parsons (21 February 1851 – 26 January 1923), apparently known for developing railways in South America.

- The Hon. Sir Charles Algernon Parsons (13 June 1854 – 11 February 1931), known for inventing the steam turbine.

In addition to his astronomical interests, Rosse served as a Member of Parliament (MP) for King's County from 1821 to 1834, president of the British Association in 1843–1844,[5] an Irish representative peer after 1845, president of the Royal Society (1848–1854), and chancellor of Trinity College, Dublin (1862–1867).

Scientific Studies

During the 1840s, he had the Leviathan of Parsonstown built, a 72-inch (6 feet/1.83 m) telescope at Birr Castle, Parsonstown, County Offaly. The 72-inch (1.8 m) telescope replaced a 36-inch (910 mm) telescope that he had built previously. He had to invent many of the techniques he used for constructing the Leviathan, both because its size was without precedent and because earlier telescope builders had guarded their secrets or had simply failed to publish their methods. Details of the metal, casting, grinding and polishing of the 3-ton 'speculum' were presented in 1844 at the Belfast Natural History Society.[6] Rosse's telescope was considered a marvellous technical and architectural achievement, and images of it were circulated widely within the British commonwealth. Building of the Leviathan began in 1842 and it was first used in 1845; regular use waited another two years, due to the Great Irish Famine. It was the world's largest telescope, in terms of aperture size, until the early 20th century. Using this telescope Rosse saw and catalogued a large number of nebulae (including a number that would later be recognised as galaxies).

Drawing of the Whirlpool Galaxy by Rosse in 1845

Lord Rosse performed astronomical studies and discovered the spiral nature of some nebulas, today known to be spiral galaxies. Rosse's telescope Leviathan was the first to reveal the spiral structure of M51, a galaxy nicknamed later as the "Whirlpool Galaxy", and his drawings of it closely resemble modern photographs.

Rosse named the Crab Nebula, based on an earlier drawing made with his older 36-inch (91 cm) telescope in which it resembled a crab. A few years later, when the 72-inch (183 cm) telescope was in service, he produced an improved drawing of considerably different appearance, but the original name continued to be used.
Lord Rosse

A main component of Rosse's nebular research was his attempt to resolve the nebular hypothesis, which posited that planets and stars were formed by gravity acting on gaseous nebulae. Rosse himself did not believe that nebulas were truly gaseous, arguing rather that they were made of such an amount of fine stars that most telescopes could not resolve them individually (that is, he considered nebulas to be stellar in nature). In 1845 Rosse and his technicians claimed to have resolved the Orion nebula into its individual stars using the Leviathan, a claim which had considerable cosmological and even philosophical implications, as at the time there was considerable debate over whether or not the universe was "evolved" (in a pre-Darwinian sense), a concept which the nebular hypothesis supported and with which Rosse disagreed strongly. Rosse's primary opponent in this was John Herschel, who used his own instruments to claim that the Orion nebula was a "true" nebula (i.e. gaseous, not stellar), and discounted Rosse's instruments as flawed (a criticism Rosse returned about Herschel's own). Eventually, neither man (nor telescope) could establish sufficiently scientific results to resolve the question (the convincing evidence for the gaseous nature of the nebula would be developed later from William Huggins's spectroscopic evidence, though it would not immediately resolve the philosophical issues).

William Parsons, 3rd Earl of Rosse

The largest telescope of the 19th century,
the Leviathan of Parsonstown.

One of Rosse's telescope admirers was Thomas Langlois Lefroy, a fellow Irish MP, who said, "The planet Jupiter, which through an ordinary glass is no larger than a good star, is seen twice as large as the moon appears to the naked eye... But the genius displayed in all the contrivances for wielding this mighty monster even surpasses the design and execution of it. The telescope weighs sixteen tons, and yet Lord Rosse raised it single-handed off its resting place, and two men with ease raised it to any height."

Lord Rosse's son published his father's findings, including the discovery of 226 NGC objects in the publication Observations of Nebulae and Clusters of Stars Made With the Six-foot and Three-foot Reflectors at Birr Castle From the Year 1848 up to the Year 1878, Scientific Transactions of the Royal Dublin Society Vol. II, 1878.

Lord Rosse's Telescopes

Lord Rosse had a variety of optical reflecting telescopes built.[2] Rosse's telescopes used cast speculum metal ground parabolically and polished.

- 15-inch (38 cm)
- 24-inch (61 cm)
- 36-inch (91 cm) (aka Rosse 3-foot telescope)
- 72-inch (180 cm) (aka Rosse 6-foot telescope or "Leviathan of Parsonstown"), started in 1842 and completed in 1845.

.

ABOUT THE AUTHOR

ROBERT STAWELL BALL

(July 1, 1840 – November 25, 1913)

Sir Robert Stawell Ball was an Irish astronomer who founded the screw theory.

He was the son of naturalist Robert Ball and Amelia Gresley Hellicar. He was born in Dublin.

Ball worked for Lord Rosse from 1865 to 1867. In 1867 he became Professor of Applied Mathematics at the Royal College of Science in Dublin. There he lectured on mechanics and published an elementary account of the science.

In 1874 Ball was appointed Royal Astronomer of Ireland and Andrews Professor of Astronomy in the University of Dublin at Dunsink Observatory.

Ball contributed to the science of kinematics by delineating the screw displacement:

When Ball and the screw theorists speak of screws they no longer mean actual cylindrical objects with helical threads cut into them but the possible motion of any body whatsoever, including that of the screw independently of the nut.

Ball's treatise *The Theory of Screws* (1876) is now in the public domain. His work on screw dynamics earned him in 1879 the

Cunningham Medal of the Royal Irish Academy.

In 1882 Popular Science Monthly carried his article "*A Glimpse through the Corridors of Time*". The following year it carried his two-part article on "*The Boundaries of Astronomy*".

Ball expounded the tides in *Time and Tide*: a Romance of the Moon In 1892 he was appointed Lowndean Professor of Astronomy and Geometry at Cambridge University at the same time becoming director of the Cambridge Observatory. He was a fellow of King's College, Cambridge.

In 1900 Cambridge University Press published *A Treatise on the Theory of Screws*. That year he also published *The Story of the Heavens*. Much in the limelight, he stood as President of the Quaternion Society. He was also President of the Mathematical Association in 1900.

In 1908 he published *A Treatise on Spherical Astronomy*, which is a textbook on astronomy starting from spherical trigonometry and the celestial sphere, considering atmospheric refraction and aberration of light, and introducing basic use of a generalised instrument.

His work *The Story of the Heavens* is mentioned in the "Ithaka" chapter of Ulysses. His lectures, articles and books (e.g. *Starland* and *The Story of the Heavens*) were mostly popular and simple in style.

He died in Cambridge and was buried at the Parish of the Ascension Burial Ground in Cambridge, with his wife Lady Francis Elizabeth Ball. Their children were: Frances Amelia, Robert Steele, William Valentine (later Sir), Mary Agnetta, Charles Rowan Hamilton, and Randall Gresley (later Colonel).

LECTURES

Ball became celebrated for his popular lectures on science. He gave an estimated 2500 lectures between 1875 and 1910 in towns and cities across Britain and Ireland. In 1892, 1898 and 1900 he was invited to deliver the Royal Institution Christmas Lecture. *Astronomy; Astronomy* and *Great Chapters from the Book of Nature.*

FEEDBACK

Now that you have read the book ...

Was it interesting?

Did you enjoy what you wanted to read?
Was there any room for improvement?

Let us know at:

http://www.diamondbooks.ca/feedback

Your feedback is highly appreciated.

Thank you!

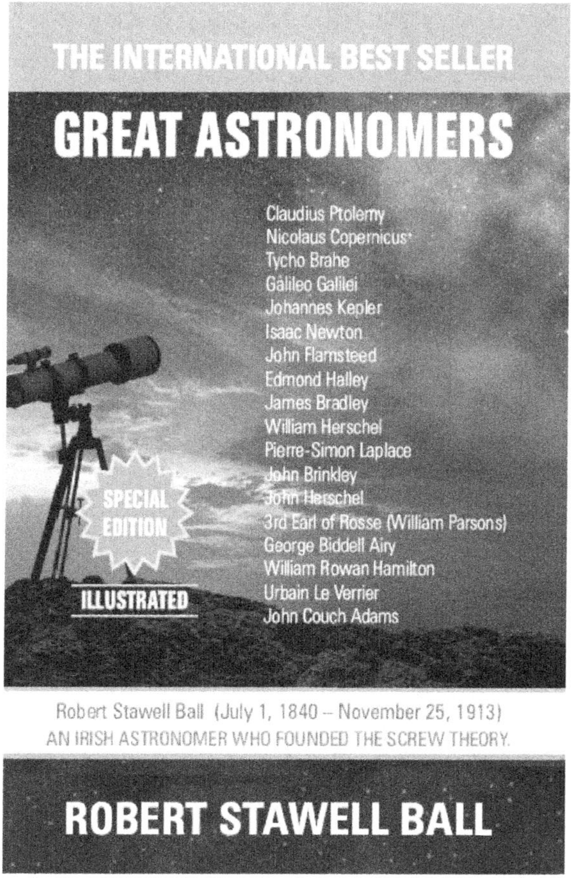
THE INTERNATIONAL BEST SELLER
GREAT ASTRONOMERS
Claudius Ptolemy
Nicolaus Copernicus
Tycho Brahe
Galileo Galilei
Johannes Kepler
Isaac Newton
John Flamsteed
Edmond Halley
James Bradley
William Herschel
Pierre-Simon Laplace
John Brinkley
John Herschel
3rd Earl of Rosse (William Parsons)
George Biddell Airy
William Rowan Hamilton
Urbain Le Verrier
John Couch Adams
SPECIAL EDITION
ILLUSTRATED
Robert Stawell Ball (July 1, 1840 – November 25, 1913)
AN IRISH ASTRONOMER WHO FOUNDED THE SCREW THEORY.
ROBERT STAWELL BALL

NOTES

NOTES

NOTES

NOTES

NOTES

NOTES